Copyright © by Harcourt, Inc.

All rights reserved. No part of this publication may be reproduced or transmitted in any form or by any means, electronic or mechanical, including photocopy, recording, or any information storage and retrieval system, without permission in writing from the publisher.

Requests for permission to make copies of any part of the work should be addressed to School Permissions and Copyrights, Harcourt, Inc., 6277 Sea Harbor Drive, Orlando, Florida 32887-6777. Fax: 407-345-2418.

HARCOURT and the Harcourt Logo are trademarks of Harcourt, Inc., registered in the United States of America and/or other jurisdictions.

Printed in Mexico

ISBN 978-0-15-362222-9

ISBN 0-15-362222-9

7 8 9 10   0908   16 15 14 13

4500449224

Visit *The Learning Site!*
www.harcourtschool.com

Working All Day

Have you ever tried to pick up a heavy item? Maybe you have had to push open a heavy door. These tasks take a lot of strength, or force. When you use force to move an object, you do **work**.

Taking a test may feel like work, but scientists don't call it work. When work is done, an object is moved. During your test, no object is moved.

Throwing a ball is fun. It's also work. The force of your arm moves the ball. Some tasks need just a little work. Others need a lot of work.

Moving a soccer ball requires work. Without work, it would not move.

This structure helps you lift, push, or pull an object. It is a kind of simple machine.

Some jobs take a lot of work. We use machines to make jobs easier. Machines change the way we do work. When you help around the house, you probably use machines. A dishwasher is a machine. A broom is a machine. A rake is a machine, too.

Some of those machines have a lot of parts, but others do not. The broom and the rake are simple machines. A **simple machine** is a tool with few or no moving parts. We use simple machines every day. We use them to lift, push, and pull objects.

 MAIN IDEA AND DETAILS How do simple machines help people?

Lifting with Levers

A lever is one type of simple machine. A **lever** is a bar that pivots, or turns, on a fixed point. A fixed point is a point that doesn't move. On a lever, the fixed point is called the **fulcrum**. To do its job, a lever must be pushed or pulled. How a lever works depends on where the fulcrum is and where the force is applied.

One type of lever has two ends and a fulcrum in the middle. Force is applied to one end. The force makes the lever turn, or pivot, on the fulcrum. One end moves. Then the other end moves. A seesaw works this way.

> **Fast Fact**
>
> What happens when a car has a flat tire? A lever comes to the rescue! The car is lifted with a lever called a jack. One end of the jack goes under the car. A person pushes on the other end. The jack lifts the car.

The fulcrum on this seesaw is in the middle.

Another type of lever has the fulcrum at one end. Force is applied to the other end. This moves an object in the middle. A wheelbarrow works this way. The wheel is the fulcrum. Force is applied when a person picks up the handles and pushes. The force moves the wheelbarrow.

A baseball bat is another type of lever. The fulcrum is at one end, and the object to be moved is at the other. Your wrist is the fulcrum. Force moves the bat, and the bat moves the ball.

MAIN IDEA AND DETAILS What is the fulcrum of a lever?

A wheelbarrow is a kind of lever. It makes work easier. You can use it to move a heavy load.

Turning Together

The wheel-and-axle is another simple machine. A **wheel-and-axle** is made of an axle connected to a wheel. When the wheel turns, the axle turns. If you have opened a door, you have used a wheel-and-axle. A rod connects the knobs of the doorknobs. The rod is the axle. The knobs are the wheels. When you turn a knob, or wheel, the axle turns, too.

Some faucets are wheel-and-axles. Suppose you need water from the bathroom sink. You might turn the sink's faucet handle. The handle is the wheel. When you turn it, a rod connected to the handle moves, too. The rod is the axle.

This knob is the wheel in this wheel-and-axle. It is a simple machine.

The energy collected from windmills can be used to make electricity.

A wheel-and-axle can be small or very large. Windmills collect energy from the wind. The arms of a windmill are the wheel. The wind's energy makes them spin. The arms are connected to axles. When the arms spin, the axles turn, which starts the machinery in the windmill.

Fast Fact

California has more windmills than any other state. In recent years, they collected enough wind power to supply almost 700,000 homes with power!

CAUSE AND EFFECT What must happen to the axle if the wheel of a wheel-and-axle turns?

Why Pull Pulleys?

A **pulley** is a simple machine made up of a wheel with a rope around it. You pull on one end of the rope, and the other end moves in a different direction. Pulleys can move very heavy objects. They can bring objects up from below the ground. They can also lift objects high into the air.

There is probably a machine that uses a pulley right outside your school. A flagpole uses a pulley system. Most flags are attached to hooks on a rope. The rope goes around a wheel at the top. A person standing on the ground pulls the rope. With each pull, the flag rises higher.

rope

wheel

Pulleys can look different, but they all have the same basic parts.

pulley

A pulley makes raising a flag much easier.

Some wells use pulleys. A bucket is needed to get water from deep inside a well. It is easy to lower an empty bucket into a well. Lifting the bucket filled with water back up is much harder. The bucket is tied to one end of a rope that wraps around a wheel. To lift the bucket, a person pulls down on the other end of the rope. The rope moves smoothly around the wheel. The pulley makes it easier to lift the bucket.

 MAIN IDEA AND DETAILS How do pulleys make work easier for people?

Ramp It Up!

An **inclined plane** is a simple machine made of a flat surface set at an angle. A ramp is an inclined plane. People can move very heavy objects with ramps. For example, ramps are used to move pianos.

Inclined planes make it easier to move up and down. A path up a hill is an inclined plane. Stairs and ladders are inclined planes, too. Slides are inclined planes. Skateboarders ride up ramps. Sometimes, simple machines can be fun.

> **Fast Fact**
> Scientists think that ancient Egyptians may have used ramps to build the pyramids—in 2560 B.C.! People moved huge stones up the ramps.

This ramp is an inclined plane. It makes moving items into the truck easier.

A **wedge** is made of two inclined planes placed back-to-back. Wedges are used to force two things apart or to split one thing into two pieces.

A wedge can look like two ramps. When we use a ramp, the ramp is still. We move the object up the ramp. When we use a wedge, it's the wedge that moves.

Many tools are wedges. An ax is a wedge, and so is a chisel. A knife is a wedge, too.

A crowbar is a tool that uses a wedge and a lever. Crowbars can be used to open stuck doors. The wedge end is placed under the door. Force is applied to the lever to open the door.

 COMPARE AND CONTRAST Compare ways in which people use ramps and wedges.

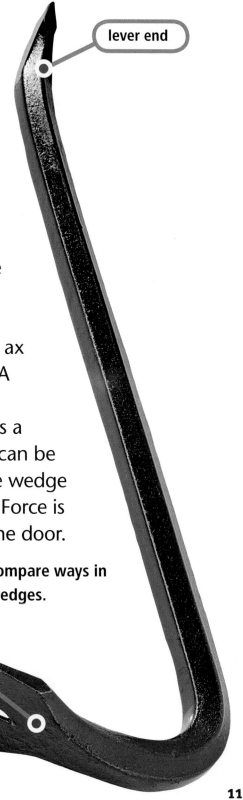

lever end

wedge-shaped head

The crowbar joins two simple machines to make a tool.

Why Use Screws?

You might not think that something small like a screw is a machine, but it is. A **screw** is a simple machine that you turn to lift an object or to hold objects together. Screws look a little like nails. Yet nails are smooth on their sides, and screws are not. Screws have threads on their sides.

Nails hold things together, but they can fall out of holes as easily as they are pushed in. Threads on screws help hold objects together. The threads also prevent the screws from falling out. Some screws have threads that are far apart. Others have threads that are close together.

The next time you open a bottle of water, look at the cap. Many bottles have caps that are screws. You turn the cap to lift it off the bottle.

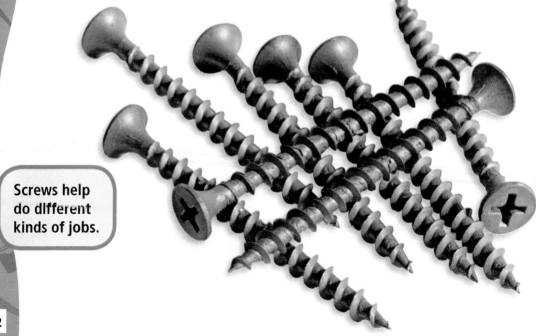

Screws help do different kinds of jobs.

Many big machines are held together with screws. Screws are used to make buildings and furniture, too. The screws must be put in very tightly so that what they are holding together doesn't move or fall apart.

COMPARE AND CONTRAST How are a nail and a screw alike? How are they different?

Fast Fact

A kind of screw for moving water was invented in ancient Greece by a scientist named Archimedes.

Screws are needed to hold this furniture together.

Inventions

Inventors make simple machines to make work easier. The modern can opener was invented in 1870. It uses more than one simple machine. The cutting wheel is a wheel-and-axle. The edge of the wheel is a wedge. The handle is a lever.

Large structures built in ancient times often used simple machines. Hundreds of years ago, inclined planes might have helped people lift huge rocks to great heights to build the pyramids of Egypt.

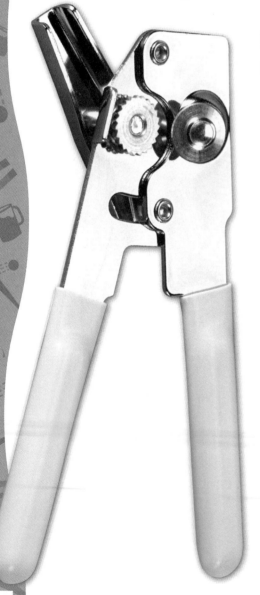

The can opener uses several simple machines—wheel-and-axle, wedge, and lever.

Discoveries sometimes require simple machines, too. Today, simple machines help scientists learn more about the pyramids. Shovels are levers with wedges at the bottom. Shovels help scientists dig up parts of pyramids buried in sand.

 COMPARE AND CONTRAST How is a rake like a can opener? How are they different?

Summary

Scientists call *work* the force to move an object. Simple machines help people do work. There are six kinds of simple machines. The lever is a bar that pivots on a fulcrum to help move objects. In a wheel-and-axle, a wheel and an axle are connected and turn together. The pulley is a wheel with a rope around it. It pulls objects up and down. Inclined planes help people move themselves or objects up or down. The wedge breaks things apart, and the screw holds things together.

This simple machine, a lever, makes work easier.

Glossary

fulcrum (FUHL•kruhm) The fixed point on a lever (4, 5, 15)

inclined plane (in•KLYND PLAYN) A simple machine that makes moving or lifting things easier (10, 11, 14, 15)

lever (LEV•er) A simple machine made up of a bar that pivots, or turns, on a fixed point (4, 5, 11, 14, 15)

pulley (PUHL•ee) A simple machine made up of a wheel with a rope around it (8, 9, 15)

screw (SKROO) A simple machine that you turn to lift an object or to hold two or more objects together (12, 13, 15)

simple machine (SIM•puhl muh•SHEEN) A tool with few or no moving parts that helps people do work (3, 4, 6, 8, 10, 11, 12, 14, 15)

wedge (WEJ) A simple machine that is made up of two inclined planes placed back-to-back (11, 14, 15)

wheel-and-axle (weel•and•AK•suhl) A simple machine made up of an axle connected to a wheel that both turn together (6, 7, 14, 15)

work (WERK) The use of a force to move an object (2, 3, 9, 14, 15)